T0166776

The Biodiversity of India

Erach Bharucha

Mapin Publishing
in association with
TATA Power Company Ltd.

ABOUT THE CD ROM

The visuals on the biodiversity of India depicted in this CD include photographs taken during my visits to 62 of our national parks and wildlife sanctuaries over the last 35 years. The pictures are small fragments of memories that are frozen in time, which document some of the thrills of wandering through the Indian wilderness. They have captured the exciting adventures that have been viewed through my camera lens. A total of 566 Protected Areas are distributed throughout India from the Himalayas to Kanyakumari and from Rajasthan to the North-east and the Andaman and Nicobar Islands. Covering only 4.6 percent of the country, they represent a small part of the wide diversity of our living legacy of wild species and wilderness areas.

Similarly, the 1,300 photographs, 400 illustrations, 5 animations, 21 video clips and 26 birdcalls that are documented in this CD-ROM, while representative of the great variety of life forms and landscapes found in the country, present only the tip of the gigantic iceberg that is India's biodiversity.

The CD-ROM caters to the information needs of children, adults and people with a special interest in fields related to conservation biology.

It is hoped that the visuals, descriptions and narrative contribute towards producing a nation-wide thrust for the conservation of India's biological diversity.

THE BIODIVERSITY OF INDIA

India's biological diversity includes a great variety of plant and animal species, their genetic variability and the organisation of species into different wilderness ecosystems. An incredible diversity of life forms is present in the wild and in the wide variety of traditional crops and livestock that people have nurtured over thousands of years in our country.

What is biodiversity? Who uses it? How is it degraded? What are the various methods for its conservation? These are essential questions that are the keys to preserving this valuable asset.

Biodiversity is the essence of future development of new pharmaceuticals, industrial products and agricultural growth. Most of these plant and animal species are found in undisturbed ecosystems in different parts of India. The "hot spots" of biodiversity are located in the evergreen forests of the Western Ghats, the North-east and the Andaman and Nicobar Islands. Equally valuable is the biodiversity in coral reefs and wetlands. Most of the plant and animal diversity of less disturbed ecosystems is now found only in our national parks and wildlife sanctuaries. These are the last refuges of this great natural heritage.

I am grateful for the support of TATA Power Company Ltd. in the production of this CD-ROM and to Jaya Rai, Sanju Kadapati, Anand Hariharan and several other friends who have helped me create this output over the last three years.

Erach Bharucha, Pune, 2001

The forests of the Andamans have evergreen f
of trees surrounded by mangroves on the seas

AN INTERACTIVE EXPERIENCE ON BIODIVERSITY

This booklet presents an overview of the issues that are covered in the CD-ROM, which are briefly described in the following paragraphs. A great proportion of the residual wilderness of India is now under great threat. Its unique landscapes are shrinking as intensive forms of agriculture and rapid urbanisation and industrial growth spread through the country.

But can this process of development last if the wilderness vanishes?

Can humanity survive without the support of Nature herself?

Modern science has serious doubts about the possibility of the long-term survival of the human race if mankind continues to degrade natural habitats. By creating an extinction spasm of enormous proportions, current development strategies will extinguish a large part of the 1.8 million known species that have diversified through millions of years of evolution. The extinction of species cannot be reversed. Future generations of India's people will hold us responsible for this great loss.

Biodiversity, which is the living aspect of Nature, is present at three levels:

- In the variety of species of plants and animals.
- In the genetic variability of individuals.
- In the ensemble of species that are organised into ecosystems.

The complexities of Nature's **ecosystems** and the interdependence of living creatures provide a perspective of the fragility of the wilderness areas of India. This includes forests, grasslands, semi-arid areas, wetlands, rivers, mountains, islands, coastal habitats and marine ecosystems.

The **species** selected for this CR-ROM represent the diversity of the large variety of plants and animals in India—their distinctive features, distribution, habits and conservation status. This provides a general survey that is only a glimpse through a keyhole at the enormous and unique diversity of life found in this country.

The lives of traditional "**ecosystem people**" who live in the wilderness still have to use Nature's resources to subsist in the country's different landscape types. They have been left out of the process of development and in fact pay the price of preserving the country's biodiversity in our national parks and sanctuaries. It is their access to subsistance resources that is being curtailed to protect wilderness habitats and wildlife.

The present effort at **conservation** of biodiversity at the national level is aimed at preserving residual habitats and critically endangered species. Efforts at conservation can only be successful if this is seen as a national priority. The key is to secure the support of the people, especially those who live alongside the wild creatures of the wilderness.

India's superb patches of relict wild areas and their great variety of living creatures make up a national heritage of enormous value to mankind.

BIODIVERSITY

Biological diversity, or biodiversity, is that part of Nature which includes the differences in genes among the individuals of a species, the variety and richness of all the species in a region as well as the various types of ecosystems within a defined area.

Life on earth is like a fragile spider's web, made of a large variety of interlinked plant and animal species. Extinction of species is akin to breaking the mooring threads of this fragile web of life. Our actions can lead to a complete breakdown of this intricate web.

The linkages between the three levels of diversity at genetic, species and ecosystem levels are crucial to maintaining life on earth.

GENETIC DIVERSITY

This is one of Nature's most wonderous features. Genes make each individual of a species different from the other. The number of permutations and combinations of genes is a special aspect of biodiversity that is used for developing new crops and in genetic engineering. This colossal variation is an important aspect of India's future economic growth and development.

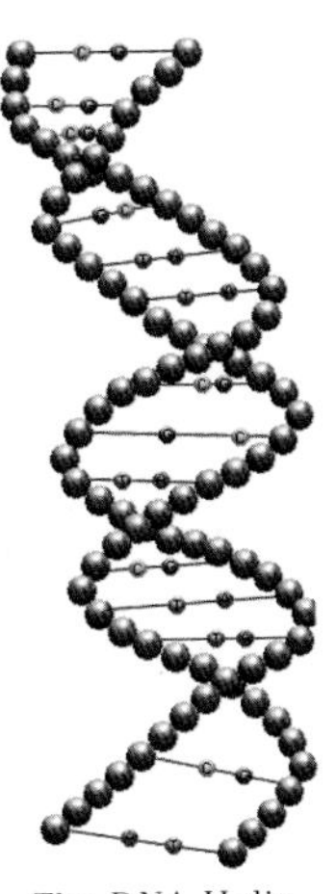

The DNA Helix

SPECIES DIVERSITY

We know there are over 1.8 million species of plants and animals on earth. The number of unknown species is much higher. It is estimated that about 50 percent of the earth's species are likely to become extinct during the next couple of decades.

There are 45 thousand plant species in India and 77 thousand animal species that have been recorded to date. This is 7 percent of the world's plant species and 6.4 percent of the global animal species, which makes India a mega diversity nation.

ECOSYSTEM DIVERSITY

The enormous variations in landscapes based on climate and topography create a diversity that varies greatly from one region to another. India has 10 distinctive biogeographic zones, which is one of the country's most distinctive features.

INDIA'S BIODIVERSITY

INDIA'S SPECIES

India is extremely rich in plant and animal species. As we continue to explore the species found in our threatened wilderness, a greater understanding of the enormous economic value embedded in the genetic makeup of these little known species emerges and gains an increasing importance. Several of these plant and animal species are found only in our country. Many are highly endangered by human population growth and unsustainable development strategies.

Species diversity and ranking of Indian fauna and flora from a global perspective

	World Ranking	Number of Species
Mammals	8th	350
Birds	8th	1,200
Reptiles	5th	453
Amphibia	15th	182
Angiosperms	15th–20th	14,500

Source: Adapted from 'Conserving the World's Biological Diversity', IUCN, WRI, CI, WWF-US, and the World Bank, 1990.

Opposite page: India's Ecosystems—the evergreen forest of the high rainfall tracts of the Western Ghats is a "hot spot" of biodiversity.

The Animal World

We are often unaware of the large diversity of animals that live in our natural habitats. While we are aware of the "glamour" species, essentially large mammals and several types of birds, we rarely appreciate the fact that there are several species of animals that we do not know about.

Vertebrates

The vertebrates include mammals, birds, reptiles, amphibia and fish. They are found on land as well as in marine and freshwater habitats.

The vertebrates include a group of mammals called amniota. They have a special membrane covering the embryo. This arrangement replaces the aquatic environment in which the larvae of fish and amphibia develop.

Elephants need wide tracts of forests for their survival.

Mammals

The blackbuck, India's only true antelope of the open plains, is now threatened by the conversion of its habitat into agricultural land.

The tiger which symbolises India's jungles is now under threat of extinction due to habitat loss and poaching for its skin and bones for use in Chinese medicine.

The wild ass lives only in the Little Rann of Kutch.

Birds

Birds of prey are important apex predators of the world of birds. These species are under threat due to the rampant increase in the use of insecticides, which they ingest through the food-chain.

The great Indian bustard is one of our most threatened grassland species.

Reptiles

The gharial, once a common species of our rivers, lakes and marshes, is now rare.

A majority of our snakes such as the vine snake are non-poisonous.

Amphibia

The Western Ghats are home to a variety of tree frogs.

Fish

A puffer fish from coastal waters.

Invertebrates

The invertebrates consist of a variety of taxa that inhabit both aquatic and terrestrial ecosystems. Together they constitute a great proportion of life on earth both in terms of the richness of species as well their abundance. The most primitive are the protozoa. Zooplankton, which include a variety of microscopic animals, form one of the bases of the food-chains in various aquatic habitats. Thousands of different kinds of animals live in the shallow seas. Worms include a great diversity of species. Molluscs are very ancient species that live in oceans as well as freshwater. Arthropods are ancient species, which include insects, spiders and scorpions. Crustacea are primitive species that inhabit different types of aquatic habitats. They include crabs and lobsters. Antonnata are a specific superclass of animals that include centipedes and millipedes.

Microscopic zooplankton are an important part of aquatic food chains.

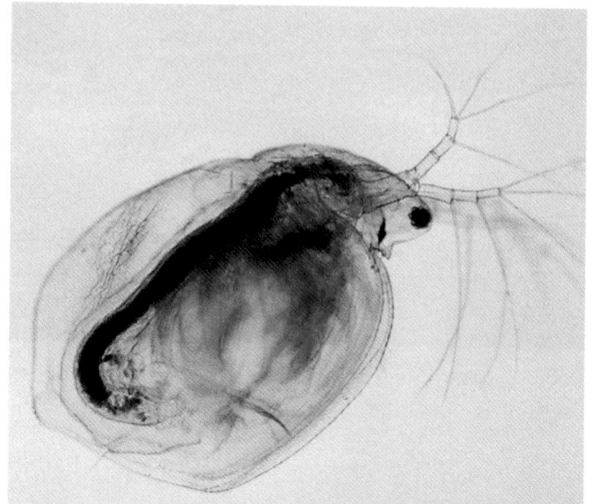

Very little is known on the diversity of molluscs.

A large variety of crabs are found in marine and freshwater ecosystems.

Insects are the most abundant species on earth. Among them, beetles have the largest number of species.

The Atlas is the largest moth in India. It is now increasingly rare.

Fungi play an important role in decomposing organic matter.

A great diversity of conebearing plants are found in the Himalayas.

The Plant World

The plant world has green plants and fungi. The kingdom of green plants, *Plantae*, has 10 phyla. The plant world can synthesise its own food material through the process of photosynthesis in which light energy from the sun is converted into chemical energy in the form of carbohydrates for growth, development and reproduction. They are thus the producers in Nature. In this process the plants release oxygen, which is a product of the chemical reaction between carbon dioxide and water that they absorb. Plants maintain the oxygen levels of the atmosphere, which is essential for respiration—a life process that supports all animals, including man. The plant world is vital for the survival of the animal kingdom.

Rhododendron trees of the Himalayas burst into bloom, dotting the landscapes with a vivid scarlet.

Orchids of great diversity are found in our forest tracts.

Habenaria is a ground orchid found in the Western Ghats.

Ceropagia is an unusual climber found in the Western Ghats.

Impatiens, a herb from Silent Valley, grows on moist rocks.

INDIA'S ECOSYSTEMS

The different ecosystems in India are seen in the 10 major biogeographic zones. Ecosystems are formed by the climatic soil and hydrological features that support species of plants and animals adapted to living in these conditions. They are also formed by thousands of food-chains that constitute the different ecosystems' webs of life. In terms of energy transfer and biomass, these food linkages and the abundance of plants, herbivorous animals and carnivorous animals can be depicted as food pyramids.

Biographic Zones of India	**%***
1. Trans-Himalayas	5.6
2. Himalayas	6.4
3. Desert	6.6
4. Semi-Arid	16.6
5. Western Ghats	4.0
6. Deccan Peninsula	42.0
7. Gangetic Plain	10.8
8. Coasts	2.5
9. North East	5.2
10. Islands	0.3

*Represents percentage of total geographical area of India.

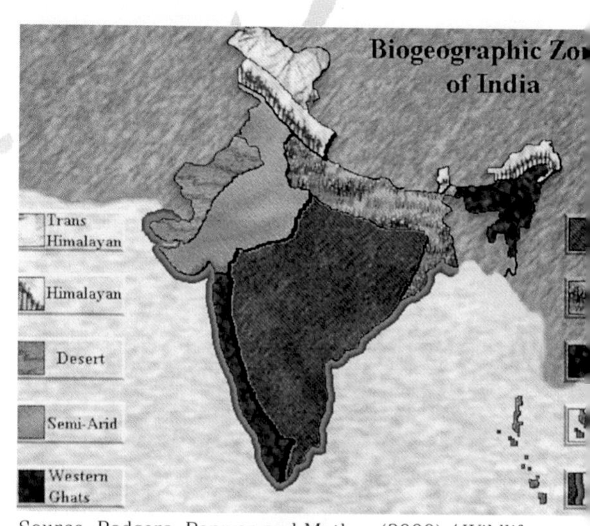

Source: Rodgers, Panwar and Mathur (2000) / Wildlife Institute of India. Map not to scale.

Forest Ecosystems

Each of the many forest ecosystems of India has its own wonderous features. They include a wide variety of plants and animals specific to each forest type.

The Himalayan coniferous forests have tall stately trees and broad-leaved forests of great beauty. They prevent soil erosion and provide a wealth of natural resources for the Himalayan people.

Coniferous forest, Himalayas.

The evergreen forests of the Western Ghats, the North-east and the Andaman and Nicobar Islands are some of the world's most species-rich areas. These are among the most ecologically fragile regions in India.

Above: Semi-evergreen forests, Western Ghats.
Below: Evergreen Shola forests of South India.

The seasonal changes in deciduous forests is a remarkable feature. These forests of mainly teak in peninsular India and sal in the plains of the North-east have been severely exploited for their timber.

The vegetation of thorn forests has evolved to cover the very arid regions of India. There are several useful xerophytic species in these forests.

Above: Deciduous forests.
Below: Thorn forests.

The Grasslands

The great open grassland vistas are being rapidly lost to agriculture and urbanisation. They include Himalayan pastures, semi-arid grasslands in the West and in the Deccan Plateau and the grasslands of the Sholas in the South.

Above: Terai grasslands.
Below: Semi-arid grasslands of the Deccan.

The Desert

The desert and semi-arid landscapes of India form a habitat in which there are highly specialised species of plants and animals.

The Thar Desert.

The Islands

The island ecosystems are rich in biodiversity and are extremely fragile. They include undisturbed forest tracts, coral reefs and the marine ecosystems that surround them.

Andaman Islands.

The Coast

The coastal belt is the home of agriculturists and fisherfolk. Fish and crustacea, when harvested sustainably, are a major source of food for the world.

The coastal ecosystem forms a habitat for shore birds.

The Marine Ecosystem

Marine ecosystems include highly diverse areas such as coral reefs, which are nearly as rich in species as evergreen forests. Sea grasses and algae are a rich source of food products and could be harvested in the future.

The marine ecosystem.

NATIONAL PARKS AND WILDLIFE SANCTUARIES

India has 566 Protected Areas, which are managed to preserve their naturalness and protect all their plant and animal species. Of these, 86 are national parks and 480 are wildlife sanctuaries. National parks have a higher conservation importance and a higher level of protection than the wildlife sanctuaries. These Protected Areas, however, cover a meagre 4.66 percent of India's landmass. Several Protected Areas are now under increasing pressures from surrounding development projects and due to the growth of human populations both within and on their periphery.

A majority of our sanctuaries have been established to preserve glamour species of mammals. Much less importance has been given to other less obvious taxa such as birds, reptiles, amphibia, fish and invertebrates. Protecting plant life has not been adequately considered in the selection of areas to be notified as national parks or sanctuaries. A large majority of our Protected Areas have been established to protect mainly forested areas, while protection of other ecosystems such as grasslands, wetlands, rivers, lakes and other unique habitats has been given insufficient priority. The Protected Area Network must provide a greater level of coverage for hot spots and relict ecosystems that cover only a small part of our country and are thus highly threatened.

Many of our species are found in only a single Protected Area. Several plant and animal species are on the brink of extinction. The Protected Areas need local support and the country needs a mass conservation movement to protect and preserve its biological diversity.

Forest sanctuaries like Corbett National Park protect a large variety of threatened plant and animal species such as the elephant.

Wetland sanctuaries support fish, crustacea and waterfowl.

Brakish water ecosystems are rich in species diversity of aquatic flora and fauna.

Himalayan Protected Areas preserve some of India's most fragile ecosystems.

The hangul is found only in Dachigam Wildlife Sanctuary in Kashmir.

The last three Siberian cranes—now seen only in Bharatpur—at the Keoladeo National Park in Rajasthan.

ECOSYSTEM PEOPLE

Traditional communities have a great deal of indigenous knowledge of the biodiversity that surrounds them. We need to learn from them.

For the last two million years of civilisation people have lived closely with Nature. This has rapidly changed only during the last few decades.

In India several tribal communities continue to practise traditional ways of life that are linked closely with the ecosystem they live in. These ecosystem people are dependent on the natural resources of wilderness systems that provide them with all their needs. India has the largest number of tribal communities in the world. These are homogenous societies in which there is little individual specialisation of work. They have lived till recently in relative isolation from other people. They have clearly defined regional distributions and have maintained their unique identities. Each tribal society has specific cultural and religious customs, which have evolved to suit the ecosystem.

Tribal people of Similipal.

Tribal children in Kaziranga, Assam.

A tribal hut made with all the resources that local people collect from the forest.

The Bhil people of the Dangs live close to Nature and are excellent fishermen.

Several communities such as the Bishnois have sustainable lifestyles and have protected Nature in India.

The Bishnoi community in Rajasthan has traditionally protected trees and the threatened blackbuck in the region.

THE LIFE SUPPORT SYSTEMS

Left: Water—the most valued resource on earth.
Below: Every drop of water counts.

Preserving ecosystems and protecting species are both vital components of India's conservation strategy.

Preserving ecosystem diversity is only possible by protecting wilderness.

Protecting endangered species such as the Malabar giant squirrel can only be done by preserving their habitat.

CONSERVATION

Conserving biodiversity is about keeping life on earth alive for posterity.

Ficus species have been protected by tradition in Indian culture. Modern science now considers them as key stone species which are of great importance in the ecosystem as they support a large variety of insects, birds and mammals that feed on them.

Conservation of wildlife is not a recent concept in India. The preservation of all forms of life both great and small has formed a part of our cultural ethos. Wildlife, especially our major glamour species, has been given varying degrees of protection over several centuries for many reasons. Thus moving from wildlife protection into the broader concept of biodiversity conservation should have been a very natural process. However, these efforts have usually been restricted to useful species of plants and animals or those that we find glamorous or interesting. Conservation of all species as part of our life support systems has not been sufficiently addressed.

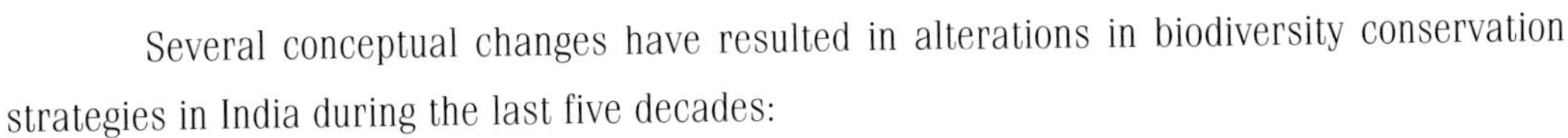

Several conceptual changes have resulted in alterations in biodiversity conservation strategies in India during the last five decades:

1950s — Importance was given to the protection of major wildlife, especially glamour species of mammals, for which notification of the first few sanctuaries and national parks was done.

Early 1960s — Importance was given to conserving habitats, mostly in forests, through notification of more sanctuaries and national parks. Changing priorities evolved from protecting "game" species to preserving their habitats.

Late 1960s — Priority was given to conserving endangered and rare species and their habitats. A new realisation grew that some of our species were on the brink of extinction.

1970s — In the face of a growing concern for habitats that had been neglected, conserving ecosystems other than forest, such as wetlands, grasslands, coasts, deserts, etc., was initiated.

1980s — Conservation of genetic diversity, including cultivars, became a growing concern.

1990s — Conservation as a broad based science, to protect and preserve all genes, species and ecosystems together, formed a new and well accepted concept.

2000 onwards — We will now be writing the history of the future of mankind on earth. Preserving biodiversity through conservation action must become a part of our strategy for caring for our earth.

WHY DO WE NEED CONSERVATION OF BIODIVERSITY?

People who live in natural ecosystems use biodiversity to support their daily needs. They also derive economic support through marketable products collected from wilderness areas.

The wilderness ecosystems support rural landscapes by preserving soil and water regimes and by maintaining microclimates. These are ecosystem services which are in turn linked with communities of plants and animals of the different wilderness ecosystems. Bees, birds and animals help pollinate crops. Undisturbed tracts of multi-storied forests are mainly responsible for maintaining the "climate near the ground"—the high moisture levels and greater humidity increase the "live" period during which the streams flow. Quite apart from the local wilderness people, nearby rural people depend on the natural ecosystems for fuel wood, fodder and other products that form basic resources on which agriculture is highly dependent. In turn, urban communities and industrial development are dependent on the agricultural sector.

Urban communities need by far the largest amounts of natural resources and energy. The industrial belts consume ever greater amounts of electrical energy. It is when wilderness ecosystems are converted into Hydel projects that such energy needs are met.

Modern man uses the species found in wilderness ecosystems as the basic material for biotechnology through which new pharmaceutical and industrial products are produced. Genetic engineering is closely linked to the wide variety of genes present in wild plants and animals. This is one of Nature's largest resources for the future development of mankind.

Exploiting wilderness and changing it into other forms of land use leads to a loss of biodiversity and extinction of species that is in fact a great national and global asset.

In the national parks and wildlife sanctuaries a variety of conflicts occur between the objectives of Protected Area management and the resource needs of local people. Animals such as elephants, wildboar, blackbuck, sambar and other wild herbivores damage crops, which causes conflict with local people. Tigers and leopards lift cattle and, occasionally, tragic human deaths occur. This loss of human life and livelihood must be compensated adequately if the nation wishes to preserve the enormous economic resource that is inherent within the biological diversity of India's wilderness.

The publication of this book and CD-ROM has been made possible with generous support from TATA Power Co. Ltd.

First published in India
in 2002 by
Mapin Publishing Pvt. Ltd.
Ahmedabad 380013 India
Tel: 91-79-755 1833 / 755 1793
Fax: 755 0955
email: mapin@icenet.net
www.mapinpub.com

Simultaneously published in the
United States of America
in 2002 by
Grantha Corporation
80 Cliffedgeway
Middletown, NJ 07701

Distributed in North America by
Antique Collectors' Club
Market Street Industrial Park
Wappingers' Falls, NY 12590
Tel: 800-252 5321
Fax: 845-297 0068
email: info@antiquecc.com
www.antiquecc.com

Distributed in the United Kingdom,
Europe and the Middle East by
Art Books International
Unit 14, Groves Business Centre,
Shipton Road,
Milton-under-Wychwood
Chipping Norton, Oxon. OX7 6JP
Tel: 01993-830000
Fax: 01993-830007
email: sales@art-bks.com
www.artbooksinternational.co.uk

Distributed in Asia by
Hemisphere Publication Services
240 MacPherson Road
#08-01 Pines Industrial Building
Singapore
Tel: 65-741 5166
Fax: 65-742 9356
email: info@hemisphere.com.sg

ISBN: 81-88204-06-4 (Mapin)
ISBN: 1-890206-40-7 (Grantha)
LC: 2001096633

Design by: Beena Hemkar /
Mapin Design Studio
Processed by Reproscan, Mumbai
Printed by Tien Wah Press,
Singapore

CD Credits

Author, Concept Design, Still and Video Photography
Erach Bharucha

Software Development
Anand Hariharan

Animation, Design, Illustrations and Development
Jayalaxmi Rai

Word Processing and Formatting
Sanjay Kadapatti

Content Experts

Botanical Information and Field Identification
Aparna Watve

Reptile and Amphibian Information and Field Identification
Ashok Captain

Insect Information, Field Identification, Testing and Feedback
Sujoy Chaudhari

Insect Information and Field Identification
Radhika Vaidya

Bird Calls and Bird Distribution
Niloufer Irani

Editing, Testing and Feedback
Statira Wadia

Behafrid Patel

Shilpi Daniel

Layout Advisor
Zehra Tyabji

Questions and Answers
Shamita Kumar

Audio

Narration
Suchet Malhotra

Music
Prem Chinmaya Dunster – Spiral Dance

Prashant Shetty

About the Author

Dr. Erach Bharucha is a Consultant Surgeon and Professor of Surgery. He is also the Director of Bharati Vidyapeeth's Institute of Environment Education and Research, which is a unique institution offering doctorate, master degree and other courses. Dr. Bharucha has travelled all over the country over the last 35 years studying the wilderness and documenting wildlife and their varied habitats through his photographs. He has been associated with conservation organisations such as Bombay Natural History Society, World Wide Fund for Nature, Salim Ali Centre for Ornithology and Natural History and Wildlife Institute of India as a member of their executive bodies and research committees. He has been principal investigator for a variety of conservation research projects, which have been sponsored by the Ministry of Environment and Forests, Government of India, with a special focus on conflict resolution between the needs of conservation and of people. Dr. Bharucha has spent a large part of his life spreading the message of Nature conservation to children and adults from various sectors of society.

This unique CD-ROM on India's biodiversity has been sponsored by TATA Power Company and is a result of the author's long association with the Tatas to bring about ecologically sensitive management to ensure conservation of the four Hydel catchments in the Mawal and Mulshi *talukas,* districts, of the Western Ghats.

Dr. Bharucha and his team have worked intensively over the last several years to put together this CD-ROM as a tool to create public support for the conservation of Nature in the residual wilderness of India and to help protect all its threatened wild creatures.